Thomas Spencer Cobbold

Tapeworms (Human Entozoa)

Their Sources, Varieties and Treatment

Thomas Spencer Cobbold

Tapeworms (Human Entozoa)
Their Sources, Varieties and Treatment

ISBN/EAN: 9783337371319

Printed in Europe, USA, Canada, Australia, Japan

Cover: Foto ©berggeist007 / pixelio.de

More available books at **www.hansebooks.com**

TAPEWORMS

(HUMAN ENTOZOA)

THEIR SOURCES, VARIETIES, AND TREATMENT

WITH ONE HUNDRED CASES

BY

T. SPENCER COBBOLD, M.D., F.R.S., F.L.S.

FORMERLY LECTURER ON PARASITIC DISEASES AT THE
MIDDLESEX HOSPITAL MEDICAL COLLEGE

THIRD EDITION

LONDON

LONGMANS, GREEN, AND CO.

1875

PREFATORY NOTE.

ESSENTIALLY practical in its character, this volume is neither intended to supersede my larger treatise on ' Entozoa,' nor my smaller volume on ' Worms ' which, in the form of Lectures, deals with the intestinal parasites generally. Almost every word of the present edition has been re-written, more than four-fifths of the text being entirely new.

The favourable reception of the previous editions by the profession has led me to believe that a short account of cases occurring in my private practice might not be unacceptable. To gain the necessary space, however, without rendering the work inconveniently large, I have been obliged to omit the illustrations (which were chiefly useful to the student), also the Appendix given in the first edition, and likewise the chapter on Threadworms given in the second edition.

The threadworms, or Ascarides, as they are sometimes erroneously termed, are fully discussed in the ' Lectures on Practical Helminthology,' above referred to.

In regard to the subject of Hydatids, briefly considered in the former editions of this book, I have also thought it best to reserve what I have to say on that head for separate publication.

T. S. C.

42 HARLEY STREET, CAVENDISH SQUARE, LONDON:
August 1875.

CONTENTS.

———◦◦———

TAPEWORMS.

INTRODUCTION.

As REGARDS the dealings of Providence in relation
to the welfare of humanity, different persons entertain remarkably opposite views. It is certainly not
easy in every case to trace a direct connection between wrongdoing and suffering; and most persons
happily succeed in persuading themselves that,
whatever inconveniences they have to put up with
here below, all is ordered for the best.

So long as they enjoy good health, and are not
troubled with internal parasites, this kind of doc-
trine is extremely agreeable to them; but when
they find themselves constituted as 'hosts' or 'en-
tertainers' of tapeworms, or other noxious forms of
entozoa, then they are apt to begin to doubt the
need-be for such an arrangement in the parasite's
favour.

The so-called fiery serpents of the wilderness
(probably nematoid entozoa), and especially the
worms which devoured King Herod, seem to assure

some people that the possession of similar 'guests'
must be regarded as furnishing special evidence of
Divine displeasure ; yet, when they think the matter over more considerately, they cannot understand
how it is that all the higher animals are liable to
the same or similar verminous disorders. The difficulty of maintaining their original and precon-
ceived opinions becomes considerably enhanced
when they further reflect upon the intimate relation
subsisting between the forms of parasites dwelling
in the human territory and those occupying the
bodies of our domestic animals.

The difficulty to which I have alluded becomes
yet further increased since it has been found, by
the light of modern scientific discovery, that certain
apparently distinct forms of entozoa, dwelling in
different bearers, are only stages of growth of one
and the same parasite, whose welfare and existence
are absolutely dependent upon the life of either
'host.' Thus, as obtains in the case of the so-
called common or pork tapeworm, the human 'host'
must devour part of the animal 'host,' so as to per-
mit of the migration of the parasite, which is
effected passively and independently of any will on
the part of the entozoon.

Paradoxical as it may seem, we human beings
are essential to the existence of particular species
of tapeworm, and it would appear that it is only by
our accepting Mr. Darwin's hypothesis of Natural

Selection that we can escape the undignified conclusion that parasites were specially created to
dwell in us, and consequently, also, that we were
predestined to entertain them.

Whatever view we may adhere to, the facts
speak for themselves. Undoubtedly the human
body, in common with the bodies of animals, con-
stitutes a peculiar territory for entozoa. To reside
in this home of theirs, and to enjoy themselves at
our expense, is their especial prerogative. The
entire organisation of these creatures is admirably
contrived for this purpose. Some will resist ex-
tremes of heat and cold. Many are furnished with
a special cyst or protective covering. Most of the
juvenile forms are supplied with a tearing or boring
apparatus; the same creatures in their adult condition having remarkable ' hold-fasts ' for the purpose
of anchorage.

For details respecting the varied and singular
arrangements which obtain in intestinal and other
internal parasites generally, I must refer the reader
to my illustrated ' Introduction to the Study of
Helminthology,' and also to my published Lectures
delivered at the Middlesex Hospital. In my
' Manual of the Entozoa of our Domesticated
Animals ' I have given a complete summary of the
facts of parasitism as severally affecting the horse,
ox, sheep, dog, pig, and cat. In the present volume I
purposely restrict myself to practical considerations

connected with the full-grown or sexually mature cestodes infesting the human body. These creatures, familiarly called tapeworms, may be enumerated as follows :—

1. THE ARMED, OR PORK TAPEWORM—*Tænia solium,* Linneus; *T. humana armata,* Brera ; or *T. lata,* Pruner.
2. THE UNARMED, OR BEEF TAPEWORM—*Tænia mediocanellata,* Küchenmeister; or *T. cucurbitina grandis saginata,* Goeze.
3. THE MUTTON TAPEWORM—*Tænia tenella,* Cobbold.
4. THE BROAD OR PIT-HEADED TAPEWORM—*Bothriocephalus latus,* Bremser; or *Tænia lata,* Pallas, *T. humana inermis,* Brera.
5. THE GREENLAND TAPEWORM—*Bothriocephalus cordatus,* Leuckart.
6. THE CRESTED TAPEWORM—*Bothriocephalus cristatus,* Davaine.
7. THE ELLIPTICALLY-JOINTED TAPEWORM—*Tænia elliptica,* Batsch ; *T. cucumerina,* Bloch ; or *T. canina,* Pallas.
8. THE MARGINED TAPEWORM—*Tænia marginata,* Batsch.
9. THE TRIPLE-CROWNED TAPEWORM—*Tænia acanthotrias,* Weinland.
10. THE SPOTTED TAPEWORM—*Tænia flavopuncta,* Weinland.
11. THE RIDGED TAPEWORM—*Tænia lophosoma,* Cobbold.
12. THE EGYPTIAN OR DWARF TAPEWORM—*Tænia nana,* Siebold ; or *T. ægyptiaca,* Bilharz.
13. THE HYDATID FORMING TAPEWORM—*Tænia echinococcus,* Von Siebold.

1. THE ARMED, OR PORK TAPEWORM.

(Tænia solium.)

FROM time immemorial this parasite has been supposed to be one of the commonest forms of entozoa liable to infest the human body; but I have shown that this conclusion is erroneous. The error has arisen from the circumstance that most of the human tapeworms closely resemble one another in their general features.

To the medical practitioner it is of considerable importance that the different forms of tapeworm should be readily recognised; but to the general reader, or to the patient who may unfortunately be infested, it is a matter of no great consequence.

To enter minutely, therefore, into the differences subsisting between the two commonest forms of human tapeworm is not my present purpose, since to render these distinctions clear it would be necessary to supply a series of illustrations. The armed tapeworm, as has now been known for many years past, is derived from pork; or rather, in other words, we obtain it by eating the so-called 'measly' flesh of swine. An impression prevails in some quarters that all human tapeworms have a similar origin; but without entering into particulars I may state, once for all, that such a notion is entirely groundless. A case recently came under my notice where a medical gentleman attributed the presence of liver hydatids in his patient to the habit of pork-eating.

Every tapeworm has its own special form of larva, and the larva itself is as distinctive as its own adult parent. Every full-grown tapeworm has an aptitude, so to say, for a certain kind of residence, and, as obtains in the species under consideration, usually confines itself to one particular kind of bearer. It may almost be said that every tape-worm has its own 'host' or 'bearer,' and consequently also that every bearer carries, or is liable to carry, his or her own tapeworm.

Very contrary, indeed, does it appear to the presumed dignity of the human species, that man should be, as it were, singled out as the legitimate home and territory of a tapeworm; yet not only is this the case, but, as before hinted, science almost teaches us to aver that, so far as this life is concerned, man appears to have been made expressly for the accommodation of certain tapeworms. At all events, without man, two apparently distinct species of tapeworm could not, it would seem, exist. The pork tapeworm has never yet been found in any 'host' save man, and the same may be said of the

species derived from beef. Of course, in making
the above statement as regards cause and effect, I
am only, for the time, adopting the ordinary mode
of teleological reasoning; but however logical the
deduction may at first sight appear, I need not say
that I can hardly bring myself to believe in the
correctness of such a conclusion.

The perfect armed tapeworm, as it is usually
presented to the observer, is a long, soft, white,
jointed animal which, when alive, elongates and
contracts itself with great facility. Though commonly spoken of as a single animal, it is in truth a
compound of many individuals. The individuals
are called ' cucurbitini,' ' zooids,' or ' proglottides,'
by scientific persons, and they are likewise occasionally termed links, or joints. By whatever name
they are called they represent so many distinct segments of the body, which when fully grown are
capable of detaching themselves and of enjoying a
free and independent existence. Very annoying it
is to the human bearer to be continually reminded
by his unwelcome ' guests ' that, for their own
pleasure or life-necessities, they desire to quit his
interior. Nor, indeed, do they oblige him by de-
parting all at once; but their habit is to wander,
solitarily and in succession, as it were, as if purposely ' to plague his very life out.' This expression
is one which is not uncommonly used by persons thus
afflicted; and I have seen one or two individuals so

emaciated by sequelæ arising from the presence of tapeworm, that life itself had in reality almost been 'plagued out.' To say the least, such individuals have wished themselves deprived of existence.

The ingestion of measly pork which is not thoroughly cooked gives rise to the formation of the armed tapeworm in the intestinal canal. The larva being set free from its capsule by the action of the gastric juice, attaches itself to the lining membrane of the bowel. For many weeks, or even months, no indication of the presence of the tapeworm is perceptible; but after the expiration of from twelve to fourteen weeks fragments of the worm may be discovered to have passed *per anum.* By this time, at least, the tapeworm reaches its adult condition, and if nothing be done to dislodge it, the worm may remain within the bearer for six, eight, ten, or even twenty years. So far as my observation extends, the worm grows comparatively slowly after five or six years' residence. Its powers for evil appear to become more and more circumscribed, but probably at least twenty-five years are necessary to bring its natural life-period to a close.

The symptoms to which the presence of tapeworm in the human body may give rise are extremely variable. Sometimes, indeed, the bearer is fortunate enough to be little inconvenienced by his guest; but even in such instances the immunity from injury is rather apparent than real. It is apparent inasmuch as the parasite gradually, and to the bearer unknowingly, steals away a portion of his health. The trifling feelings of weariness and lassitude are usually set down to other causes, and it is only when these indications are succeeded by restlessness, nervous irritability, and headache, that proper attention is paid to the true source of these symptoms. Rather than let any fellow-creature know the fact of their having tapeworm, many persons will not only endure some of these troubles, but even submit to the still more annoying inconveniences arising from the constant passage of the worm-segments. So revolting is it to human nature, especially to refined and educated minds, to be called upon to entertain the presence of such creatures, that only the gravest sufferings will induce some people to obtain medical advice. This applies not only in the case of tapeworms, but also to many other human parasites. Happily, but few individuals are *dangerously* affected by these unwelcome guests.

In bad cases the foregoing symptoms become greatly aggravated. The headache is much increased, and often accompanied with vertigo or giddiness. The sight and hearing may be affected. Noises in the head, itchings at the nose and arms, obscure pains about the body and limbs, loss of appetite, and other dyspeptic symptoms show themselves in greater or less degree in different cases. One of the most common symptoms, however, which I have noticed is the tendency to faintness. This is sometimes so marked as to create much alarm, and a person uninformed as to the true cause of the disorder might be led to treat the symptom as arising from a totally different source. In female patients, the nervous symptoms display features more or less peculiar to the sex. The restlessness and anxiety is excessive, and at times accompanied with chorea and fits of hysteria. Paralysis is not unfrequent, and I have witnessed many instances of paraplegia and hemiplegia. In the worst cases, in both sexes, the cerebral disturbance may show itself in convulsions and epileptiform seizures. I regret to have to add that in a few instances even mania itself has been entirely attributable to the presence of tapeworms in the intestinal canal.

In all severe cases the evacuation of the worm is commonly attended with perfect restoration to health ; but in those instances where the worm re-developes itself, the return of the malady may be expected. It by no means follows, however, that the identical symptoms will again show themselves, although, in the less formidable attacks, this not unfrequently takes place.

2. THE UNARMED, OR BEEF TAPEWORM.

(*Tænia mediocanellata.*)

IF I were to speak of this parasite as the common tapeworm, the majority of people would suppose that I was referring to the species previously described; and yet the facts of the case would warrant me in styling this worm the most common of all tæniæ liable to invade the human body. In general appearance it is very similar to the armed form, at least when viewed by the naked eye. Commonly it is a larger and broader animal, being at the same time rather stouter. It varies usually from fifteen to twenty-three feet in length, but specimens have been described as attaining upwards of thirty feet. It is called the unarmed tapeworm in consequence of the absence of any coronet of hooks on the head, and consequently, also, from there being no prominent rostellum or proboscis. The place of the last-named structure, however, is supplied by a small rudimentary disk, which I have seen protruded on pressure. Usually this disk forms a more or less conspicuous cup-shaped circular depression, which has been compared to and described as a fifth sucker. That it is not, in any structural sense, comparable to the true suckers, I have had abundant opportunity of ascertaining ; nevertheless I do not doubt that it is

to a slight extent capable of being used by the parasite as a supernumerary holdfast. The anchorage
thus secured, however, is by no means equal to that
obtained by the armed species, a circumstance
which explains the comparative difficulty we find in
procuring a specimen of the armed species with the
head attached.

The experimental researches long ago conducted
by Leuckart and Mosler abroad, by myself in this
country, and quite recently also by Prof. St. Cyr
in France, have satisfactorily determined the origin
of this parasite. We have together incontestably
proved that the human body becomes infested in
consequence of the consumption of veal and beef.
It may seem strange to many that we should speak
of measly veal and beef; nevertheless, it is probable
that more diseased beef exists in this country than
similarly affected pork. I mean to say that the
flesh of cattle used as food is more commonly infested with the larvæ of tapeworms than is the
flesh of swine; but the larvæ in the one case are
essentially different from those in the other.

---

3. THE MUTTON TAPEWORM.

(*Tænia tenella.*)

I HAVE in my collection several examples of a very
delicate human tapeworm, which I believe to be

referable to an armed cestode derived from the ingestion of mutton affected with measles (*Cysti-cercus ovis*). The uterine branches are more widely apart than they are in the pork tapeworm, and the proglottides are smaller and narrower. In the year 1865 I detected the presence of cysticerci in mutton, since which time I have not only repeatedly seen these cestode larvæ in the flesh of sheep, but others have confirmed the truth of my original discovery. It has thus been placed beyond a doubt that all the ordinary meats sold in the market are liable to become measled. At one time it was thought that the flesh of swine alone was liable to infection, but now we may encounter also measly beef, veal, and mutton. In favour of hippophagy it must be said that no one has yet discovered measles in the flesh of the horse.

At a meeting of the Pathological Society of London, held in the month of April 1866, I mentioned that on three separate occasions I had noticed measles in 'joints' of mutton sent to my own table. These were all, I believe, from the shoulder, but in the month of September 1874, I removed some from a 'leg' whilst at dinner. All the above-mentioned specimens of measles were in a more or less advanced condition of calcareous degeneration; yet, if no other data had been forthcoming, the case was clearly proved. My friend Dr. Kirk, H.M. Consul at Zanzibar, has stated to me that he fre-

quently encountered similar appearances. In the
year 1866, Mr. Heisch, Professor of Chemistry, obtained a perfect cysticercus from the interior of a
mutton chop. This he was good enough to place at
my disposal for the purpose of scientific description.

The mutton measle is somewhat smaller than
the pork measle, being armed with a double crown
of twenty-six large and well-developed hooks. The
larger hooks measure 1-160th of an inch in length,
the suckers having a breadth of about 1-100th of
an inch.

As I had only examined one tolerably perfect
specimen of the mutton measle (the caudal vesicle
in Mr. Heisch's example being injured), I was very
glad to see it stated in 'Nature' that Dr. Maddox
had also found an encysted bladder-worm in the
neck of a sheep. This 'find' was reported in the
journal for May 15, 1873 (p. 59). In his communication the author stated that 'the presence of
immature ova was particularly noted.'

As the notion of the existence of eggs in larval
cestodes was altogether at variance with what we
know of the phenomena of tapeworm life, I ventured
shortly afterwards (in the pages of the 'London
Medical Record' for August 6, 1873) to suggest
that the author must have mistaken the egg-shaped
calcareous corpuscles (which I had noticed so abundantly in my own specimens) for the ova in question.
I should mention that Dr. Maddox's paper was

subsequently published at greater length in the
'Monthly Micr. Journal' for June 1873 (p. 246);
the views to which I have referred being reite-
rated with confidence. In the interests of truth
I felt bound to characterise certain of the conclusions arrived at by Dr. Maddox as simply 'in-
credible,' but being most anxious to do justice to
the confirmation of previous discovery, I also stated,
what is perfectly true, that the memoir in question
'formed an important contribution to our knowledge
of the structure of the mutton measle.' I had no
idea that in pointing to mere 'errors of interpreta-
tion' I should grievously offend the excellent author
of so valuable a communication. However, very
shortly afterwards a long letter appeared in the
'Medical Record,' in which the author showed that
he was much vexed that I should have 'impugned'
the 'accuracy of his conclusions.' He defends his
position, and, if I understand his meaning correctly,
Dr. Macdonald, F.R.S., is represented as supporting
his views. Dr. Maddox says: 'We were quite
alive to the anomalous position. Hence the excep-
tionability of the case rests on more than my own
evidence.' Here I will only add that there are few
observers in whose opinion I should naturally place
more confidence than in those of the distinguished
Assistant Professor of Naval Hygiene at the Victoria
Hospital, Netley. I have little doubt that these
eminent observers will eventually yield the point in

question; but whether they do so or not, the value of Dr. Maddox's contribution as a record of the facts observed is quite independent of the question as to whether a bladder-worm or sexually immature cestode is capable of generating or containing the genuine eggs of a tapeworm.

I have given a figure of the head of this parasite in the Supplement to my 'Introduction;' but, for minute details, the excellent plate accompanying Dr. Maddox's paper should be consulted. The only specimens known to exist are contained in the collections of the various persons above mentioned.

An interesting problem remains to be solved in connection with the mutton measle. Whence does it come? In reply, I can only express the conviction that, as obtains in the case of the beef and pork measle, man himself will turn out to be the sole bearer of the adult tapeworm representative. In the absence of experimental proof I have indeed already obtained a certain kind of evidence, which, though by no means conclusive, seems to substantiate this view. As the joints of this worm bear a closer relation to those of the pork tapeworm than they do to those of the beef tapeworm, and as the former is armed, whilst the latter is destitute of hooks; and as, moreover, with one exception, all the human tapeworms have originated from cysticerci in the flesh of vertebrated intermediary bearers, it thus becomes a matter of high probability that the mutton measle is the scolcciform stage of growth of my *Tænia tenella.*

Should a fresh example of this comparatively small and slender human tapeworm present itself, I shall endeavour to obtain experimental proof of this supposed relationship by feeding a sheep with the sexually ripe proglottides. The reverse experiment is not one that could well be performed in this country, since few persons, however enthusiastic in the cause of helminthology, would care to swallow the armed cysticerci derived from the uncooked flesh of sheep.

---

4. THE BROAD, OR PIT-HEADED TAPEWORM.

(*Bothriocephalus latus.*)

THIS species, though seldom seen in England, is sometimes brought hither by persons who have been residing for a time in foreign countries. Curiously enough, however, it is indigenous in Ireland; though, as compared with the two former species, it is by no means common. It has been called the Irish tapeworm, but is much better known as the Swiss or Russian tapeworm. It is especially prevalent in Russia and Switzerland, being likewise a native of other parts of Europe, more particularly of Sweden and Germany. The disease is endemic in

c

the countries bordering the shores of the Gulf of
Bothnia.

The broad tapeworm is readily distinguishable
from the other species.   Its remarkable breadth,
associated with extremely numerous and closely-
packed joints, having a very small vertical diameter,
is alone sufficiently distinctive.   A full-grown specimen attains the length of twenty-five feet, and may
carry no less than four thousand segments.   It has
generally a more or less strongly marked brownish-
yellow tint, due to the presence of coloured eggs in
the interior of the segments.   The reproductive
apertures, instead of being placed at the margin of
the joints, are situated at the centre of each successive segment on the ventral aspect of the body.

Unlike the ordinary tapeworms, the joints of
this animal do not naturally separate themselves so
as to become independent organisms.   This circumstance is highly favourable to the patient, who is
thus spared the continual annoyance usually arising
from the daily passage of worm joints.   On the
other hand, the presence of so formidable a parasite
is seldom unproductive of disagreeable effects.

The head of the worm is perhaps even more
distinctive, specifically, than any other part.   It is
somewhat flattened from before backwards, having
two long, slit-like depressions at the sides, which by
means of muscular action afford a tolerably efficient
anchorage.

The source and development of this parasite are
points of considerable interest. The eggs are of
comparatively large size, and after expulsion and
immersion in water give passage to beautifully
ciliated embryos, which latter again give birth, as
it were, to larvæ furnished with a boring apparatus.
This consists of six hooks fashioned after the manner
of those existing in the mature eggs of other tapeworms. In what animals the larvæ subsequently
develop themselves is not ascertained with certainty,
but it seems probable that persons become infested
themselves by eating certain kinds of imperfectly
cooked fresh-water fish. Leuckart has suggested
that the intermediary bearers are probably species
of the salmon and trout family. Dr. Knoch, of St.
Petersburg, seems to think that there is no need
of the intermediate host. He believed that he had
succeeded in rearing young broad tapeworms in the
intestines of dogs. In this he was mistaken.

The symptoms occasioned by this parasite do not
differ much from those produced by the foregoing
species. According to Odier, as quoted by Davaine,
there is not unfrequently a tumid condition of the
abdomen, with sickness, giddiness, and various hysterical phenomena occurring at night. Pain in the
region of the heart, palpitations, and faintness are
also mentioned.

5. THE GREENLAND TAPEWORM.

(*Bothriocephalus cordatus.*)

THIS species, though comparatively new to science, may nevertheless turn out to be identical with a worm long ago described by Pallas and Linneus. At present it is only known for certain to infest the residents of North Greenland, but it is by no means improbable that its area of distribution may be found to embrace the regions of the north generally. It is, as it were, a sort of miniature representative of the species last described. It attains the length of about one foot, and has a small heart-shaped head, whose apex is directed forwards. The neck is so obscure that it may be said to be altogether wanting, the segmentation of the body being well marked immediately below the head.

Though so small a species, Leuckart, who first described it, counted between six and seven hundred joints. As in the broad tapeworm, the reproductive orifices are serially disposed along the centre of the ventral line, but a close inspection of the organs themselves shows that the foldings of the egg-bearing organ are comparatively more numerous. The worm is therefore very readily distinguishable from all other known tapeworms.

This entozoon does not appear to be a very

frequent resident in the human body, though it is by no means uncommon in the dog. Possibly it may yet be found in the inhabitants of some of our northern and western islands. From the smallness of its size, it can hardly occasion much inconvenience to its bearers, whether human or canine; nevertheless it exists in the dog, sometimes in very considerable numbers.

My private collection contains some preparations of this parasite, for which I am indebted to Professor Leuckart.

6. THE CRESTED TAPEWORM.

(*Bothriocephalus cristatus.*)

THIS worm, which measures between nine and ten feet in length, is, according to the discoverer, Dr. Davaine, characterised by the presence of two remarkable crest-like prominences, which together form a sort of rostellum or proboscis, covered by numerous minute papillæ. The full-grown segments are less than half an inch in breadth; the body of the parasite being much narrower than that of the broad species.

The worm is figured by Dr. Davaine in his recent article on the cestodes in the 'Dictionnaire Encyclopédique des Sciences Médicales.'

I may here remark that in addition to the two
specimens on which Davaine's description is founded,
there is (in the museum attached to the Westmin-
ster Hospital Medical College) a preparation containing several examples of human Bothriocephali.
These I believe to be referable to the crested tapeworm.

---

7. THE ELLIPTICALLY-JOINTED TAPEWORM.

(*Tænia elliptica.*)

THIS parasite is readily recognised, not merely by
its delicate form and small size generally, but also
by the circumstance of its supporting two sets of
reproductive organs in each mature joint—these
structures communicating with separate outlets
which are situated at the centre of the lateral margin of the segment.

The elliptic tapeworm ordinarily infests the cat,
but there is reason to believe that it is identical
with the *Tænia canina* or *T. cucumerina*, so common in the dog. At all events, from the evidence
put forth by Eschricht, seconded as it is by Leuckart, there is every reason for believing that one or
other of these closely allied forms (be they identical
or not) is liable to infest the human body.

It was originally stated by Eschricht that he

had received a *Tænia canina* which had been passed
by a negro slave at St. Thomas, Antilles. Probably
the species is very rare in the human body, and
possibly may only occur in the negro race. So
delicate a worm, unless present in very considerable numbers, would not be likely to occasion any
bad symptoms; hence also its presence would often
either be overlooked or disregarded.

---

8. THE MARGINED TAPEWORM.

(*Tænia marginata.*)

I HAVE no positive evidence of the occurrence of
this parasite in its adult condition in the human
bearer; but there is a tapeworm in the Edinburgh
Anatomical Museum referable to this species, which
(according to the late Professor Goodsir's assistant,
Mr. John Arthur) had been obtained from the
human body. The worm, in its full-grown state, is
common enough in the dog; but, as remarked in
my larger treatise, the principal evidence demon-
strating the occurrence of its larval representative
(*Cysticercus tenuicollis*) in man rests upon two cases
recorded in Schleissner's *Nosography* of Ireland.
One of these alleged instances, however, has been
proved by Küchenmeister and Krabbe to be that of

an echinococcus; so that, after all, there only remains the solitary case observed by Schleissner himself in which the parasite can fairly be considered as the 'slender-necked hydatid.'

To the above, however, may probably be added a specimen preserved in the Anatomical Collection at King's College, London. It was found connected with an ovarian cyst. Not improbably, examples of this cystic worm may at various times have been mistaken for small hydatids or common acephalocysts.

9. THE TRIPLE-CROWNED TAPEWORM.

(*Tænia acanthotrias.*)

THIS entozoon is at present unknown in the adult condition; its existence as a true species being based on the circumstance of the occurrence in the human body of a cestode larva having a head furnished with three rows of hooks.

Dr. Weinland, of Frankfort, when visiting America in 1858, examined a specimen of supposed *Cysticercus cellulosæ* preserved in the collection of the Boston Society for Medical Improvement. This parasite was taken from a woman about fifty years of age, who died of phthisis, being afterwards a dissecting-room subject at Richmond, Virginia. About a dozen or fifteen of the cysts were found in the cellular membrane of the muscles, and in the integuments, besides one which hung free from the inner surface of the dura mater, near the *crista galli*. In the same subject there were also numerous specimens of *Trichina spiralis*. The specimen was presented by Dr. Jeffries Wyman, and an account of the case was first published in 1857.

If subsequent discoveries should show that this parasite is only a variety of the *Tænia solium*, the mere existence of such an abnormal condition of the so-called common tapeworm is in itself a very interesting circumstance.

10. THE SPOTTED TAPEWORM.

(*Tænia flavopuncta.*)

THE discovery of this interesting little tapeworm is also due to the investigations of Dr. Weinland. In Dr. J. B. S. Jackson's catalogue of the collection of the Boston Medical Improvement Society, an account of the contents of a phial is recorded as follows:—' Specimen of Bothriocephalus, three feet in length, and from half a line to one line and a quarter in width; from an infant. The joints are very regular, except at one extremity, where they approach the triangular form, are very delicate, and but slightly connected, as shown in a drawing by Dr. Wyman.' It is further stated that the infant was nineteen months old, and that the worm was discharged without medicine, its presence having never been suspected. It was presented by Dr. Ezra Palmer, in the year 1842.

On examining the above-described fragments, Dr. Weinland found, instead of a solitary specimen, at least six different tapeworms; all of them being referable to a totally distinct and hitherto undescribed species. Unfortunately, none of the heads were present in the phial; nevertheless, it was ascertained that the worms varied from eight to twelve inches in length, the joints or segments being very broad laterally, and at the same time correspondingly narrow from above downwards. It has been named ' the spotted tapeworm,' in consequence of the presence of yellow spots lying near the middle line in each successive joint. They represent the male reproductive organs. The genital orifices are serially disposed all along one side of the worm at the margin. The eggs, for the most part, resemble those of ordinary tapeworms.

11. THE RIDGED TAPEWORM.

(Tænia lophosoma.)

IN the museum of the Middlesex Hospital there is a tapeworm which when complete must have measured about eight feet in length. It is characterised by the presence of a ridge extending throughout the entire length of the body.

The comparatively small size of the segments, and the uniform disposition of the reproductive papillæ on one side, show that this parasite is distinct from *Tænia solium*, and from *T. mediocanellata*, whilst the length of the worm, as contrasted with the foregoing species, shows that it cannot be referred to Weinland's *Tænia flavopuncta*. It is a distinct form, and even if a variety only, is remarkably divergent from the ordinary species. I have provisionally named it *Tænia lophosoma*, to indicate the presence of the crest or ridge extending the whole length of the body. Küchenmeister's ridged variety of *Tænia solium* or *T. mediocanellata*, from the Cape of Good Hope, showed an alternation of the papillæ which does not exist in this specimen.

12. THE EGYPTIAN OR DWARF TAPEWORM.

(Tænia nana.)

So far as I am aware, there is but one solitary instance on record of the occurrence of this minute tapeworm in the human body; and up to the present time we have no evidence of its having existed in any other host. It was discovered by Dr. Bilharz of Cairo at the post-mortem examination of a boy who died from inflammation of the cerebral membranes. Prodigious numbers existed. The largest specimens measured only one inch in length.

To the naked eye these worms resemble short threads, and consequently they might very readily be overlooked. The head is broad and furnished with a formidable rostellum armed with a crown of hooks. These hooks have large anterior root-processes, which, extending unusually forward, impart to the individual hooks a bifid character. The general structure, however, is essentially the same as obtains in other tapeworms.

The cysticercal source of this cestode is at present unknown.

Though interesting physiologically, it is not likely that this parasite will ever become a formidable enemy to the welfare of the human race.

13. THE HYDATID-FORMING TAPEWORM.

(*Tænia echinococcus.*)

THIS remarkably minute parasite, though not resident in man in its adult condition, is nevertheless, in one of its larval stages, of frequent occurrence in the human body. Whilst the full-grown creature seldom attains the fourth of an inch in length, the larvæ, on the other hand, acquire a prodigious size. The latter are familiarly known to the profession under the name of *hydatids*. The tapeworm itself resides in the intestines of the dog and wolf; and it is from this source that we become infested.

Much space might be devoted to an explanation of the various possible ways in which the little tapeworm eggs and their embryos gain access to the human body; but, for all practical purposes, it is here sufficient to remark that they commonly enter the mouth along with food and drink, more particularly with the latter. They may also be transported as dust, by wind and other agencies, and thus be carried directly into the mouth; or they may be brushed against the lips, adhering there for a time before being swallowed. Residence in densely populated districts, where dogs at the same time abound, is eminently favourable to the

introduction of these creatures into the human
body. Thus, though not unfrequently attacking
the wealthy, they are very much more common
amongst the poor. The frequency of the echinococcus disease in Iceland and Australia is well known,
but it would be out of place to discuss this subject
in the present volume. (See my lecture on ' Hydatids' in the 'Lancet' for June 19, 1875; where I
have given a tabulated analysis, showing the distribution of hydatids in the different organs of the
body in 700 cases collected by Davaine and myself.)

TREATMENT.

ALMOST every fresh case of tapeworm offers some
new fact, or some novel phase of previously ascertained facts that have a practical bearing on the
right management of the affection. The conceptions
of professional duty, however, as ordinarily expressed and maintained in this connection are of a
kind that tend to reduce the physician's functions
to the vulgar level of the ancient ' worm-doctor.'
To be sure, it needs some moral courage on the
part of any medical man who desires by his conduct
to rescue this department of practice from the bane-
ful influences of prejudice and quackery.

Conscious of the high order of scientific work
necessary for the advancement of practical helminthology, I have been not a little amused at the
positive terror displayed by one or two of my
medical friends lest by the publication of their
experiences about tapeworms and threadworms they
should come to be dishonoured with the title of
' worm-doctor.' For my part, it appears that the
study of any cause—let it be a worm or an in-
fectious poison of the vilest character—operating

to produce serious mischief in the human frame ought not to be considered as beneath the dignity of those whose express function it is to ʻheal all manner of disease.' The childish fanaticism to which I have alluded will not be overcome just yet; but it will eventually be stamped out.

In the pages of the ʻLancet' for 1874 (vol i. p. 793) I gave an analysis of the experiences gathered from eighty consecutive and unselected cases of tapeworm occurring in my private practice. That communication, whilst it formed a small contribution towards our knowledge of the statistics of tapeworm, was intended more particularly to show the need there is of greater care and precision in the diagnosis and management of this common disorder. All the cases there analysed were not such as would be called genuine; but the expression ʻcases of tapeworm,' was not a misnomer, inasmuch as although some of the patients had never harboured tapeworm, yet one and all of them were actually brought under my notice as genuine tapeworm cases. In this fact of the absence of the worm in so many instances where it was honestly believed to exist a lesson of great importance lies concealed. Now, considering that all of the patients concerned in these as well as in other cases that have come under my professional care had previously undergone treatment, I deem it not unfair to ask such a question as the following :—ʻ If the successful

treatment of tapeworm be so simple a matter as some
persons seem to think, how does it happen that many
patients undergo years of drugging without being
permanently cured?' Of course the answer is to
the effect that the head of the worm had not been
dislodged in any instance, and therefore the parasite
continued to grow until it again arrived at maturity.
Precisely so. I have admitted all along that there
are cases of difficulty, particularly in the treatment
of the armed pork tapeworm, but it is just these
obstinate cases which demonstrate the necessity for
special knowledge and tact in their management.

Again, as affording proof of the truth of my
opening statement, let anyone not unfamiliar with
our recent advances in experimental helminthology
take the trouble to read the discussion on the treatment of tapeworm which took place at a meeting of
the Société de Thérapeutique on June 10, 1874.
('Bullet. Gén. de Thér.') From a scientific point of
view, the want of knowledge exhibited by several of
the speakers seems to be altogether lamentable.
Thus, for example, one gentleman (M. Trasbot)
actually asserts, or is represented as having asserted,
that the flesh of the ox 'does not contain cysticerci,'
although for many years past we have recognised
beef to be the most frequent source of tapeworm
both in this and in several other countries. I may
also state, for M. Trasbot's edification, that for
many years past I have been in the habit of ex-

hibiting beef and veal measles to scores of students
and other persons interested in the matter. Not
only so, the experimental proof of the possibility of
the occurrence of cysticerci in beef and veal dates
back as far as the year 1861 (Leuckart, ' Die
Mensch. Par.' s. 406), whilst Mosler's limited experiences were announced shortly afterwards. My
own much more extended verifications were first
made known in the ' Lancet' of February 25, 1865 ;
being repeated in a more complete and emphatic
manner in the same journal during the following
August. Subsequently, in the ' Pathological Society's Transactions,' and elsewhere (1866), further
confirmatory researches were made public, some of
these throwing light upon subsidiary questions in
helminthology. Quite recently, a French experi-
menter has at length accomplished a similar result.
Professor St. Cyr, however (like M. Trasbot),
appears to have been totally unaware of the fact
of the German and English experiments, although
they were conducted ten or more years previous to
his own. This evidence of defective information
on the part of those who ought to have known better
is exceedingly discouraging ; and I wish I could
bring myself to believe that such deficiencies were
exclusively the prerogative of therapeutists and
professional men on the other side of the channel.
Certainly, it ought to be generally known that
the distinguished representative of helminthology in

Paris, M. Davaine, cannot be held to blame on
this score, since, in his recent article on the
cestodes in the 'Dictionnaire Encyclopédique des
Sciences Médicales,' he has given an admirable
summary of Leuckart's and Mosler's researches, as
well as of the experiments conducted by myself in
England.

The analysis of the 80 cases above referred
to, inasmuch as they were consecutive ones, fur-
nishes results too practical and important to be
passed over. In the present volume I shall give
brief notes of about 70 of the cases that were
thus analysed, adding 30 new ones; these latter
being also selected from cases occurring in my
private practice.

In the first place, I have to observe that 27 of
the 80 cases were those of females, the remaining
53 being those of males. This proportion of one-
third of the former to two-thirds of the latter cor-
responds, I suspect, pretty accurately with the actual
relative prevalence of the disorder in the two sexes
respectively—at all events, in tapeworm as it affects
the wealthier classes in this country. Out of 64
of the undoubtedly genuine cases, I find the proportion to be 20 to 44. Speaking generally, I think
this result is sufficiently explained by the more cau-
tious and fastidious habits of the female sex, as contrasted with males, in relation to the ingestion of
underdone meat.

A more interesting statistical feature is that
appertaining to age. Of the 80 patients, 62 were
above twenty years old, but not one of them, so far
as I could ascertain, exceeded sixty years. About
two-thirds of their ages ranged between twenty and
forty years. As regards the juvenile sufferers, 8
were under eight years of age, whilst the youngest
child must have contracted the disorder at the age
of fourteen months, and therefore at the time of its
earliest attempts to swallow flesh-food. Three others
became infested under three years of age.

In reference to the question of residence or
locality, it is interesting to notice that no less than
25 patients had contracted their disease abroad.
With one exception, all of these patients had actually harboured tapeworm at some time or other, and,
therefore, deducting this one and five other similar
delusion cases from the 80, it shows that somewhat
less than two-thirds of the genuine cases were of
English origin. Of the 25 cases from foreign parts,
12 were from India and Ceylon, 4 from the United
States, and 3 from France. The other cases afforded separate instances, where the worm appeared
to have been contracted in Peru, China, Turkey,
Egypt, Norway, Germany, and Spain. In the home
cases I include several from Ireland and Scotland.

Of larger import is the question as to the length
of time the patients had already played the part of
'host' when I first saw them, or when their cases

were originally brought under my notice by correspondence or otherwise. The period of 'entertainment' of the 'guests' varied exceedingly. I
ascertained the duration of these periods with toler-
able accuracy in 58 genuine cases. Thus 2 bearers
or hosts had passed joints for two and six months
respectively, 11 throughout periods varying from
six to twelve months, 15 from one to two years, 14
from two to six years, and 7 from six to sixteen
years. Nine other patients, not included in this
reckoning, spoke of having harboured the worm for
' several years ;' consequently it is not unfair to
suppose that the whole of these 9 might be added
to the 14 already mentioned as having carried their
guests about with them from two to six years. Be
that as it may, I find that out of the entire 58 patients referred to, no less than 32 had been suffering
more or less throughout periods varying from one to
ten years respectively, whilst three others had harboured the parasite for considerably longer periods.
The longest ' period of entertainment' by the host
was upwards of sixteen years ; but, through one of
my patients I heard of a case where the worm was
believed to have been harboured for twenty years.
From other considerations, elsewhere advanced, I
incline to the belief that the natural life epoch of an
ordinary tapeworm does not greatly exceed this
last-mentioned period.

It is scarcely worth while to offer any details

respecting either the status or occupation of the various human bearers. Suffice it to say that, after deducting the cases of females and children, I find that the majority of the patients were well-to-do persons in business. In this category travelling merchants figure prominently; but many different kinds of occupation were represented. Thus I find in the list, bankers, brewers, engineers, wine merchants, printers, clerks of every grade, butchers, grocers, &c. Nor, as the facts imply by inference, are members of the liberal professions wholly free from the propensity for eating imperfectly cooked meat. According to this record, I find that the clergy especially are apt to play the *rôle* of tapeworm bearers. Amongst military men the disorder is still more common; and, as will naturally be expected, it occurs principally in those on foreign service. The majority of the cases from eastern parts were those of officers in Her Majesty's Indian Army.

Under the category of delusive cases I place not only those patients who never had the privilege of playing the part of ' host,' but also those who, at the time they presented themselves, had actually parted company with their previously entertained guest or guests. Only in two or three instances were two or more tapeworms present in any one bearer. Altogether the delusive cases numbered 24. Of these it was clear that 6 had never had

tapeworm at any time. Thirteen had certainly
harboured tapeworms, but had got rid of their
guests without being convinced or made aware
of the fact; whilst, as regards the remaining five,
the evidence was not entirely conclusive. Anyhow,
they had all previously undergone treatment for
tapeworm; and thus I am in a position to say that
more than one-fourth of the whole 80 so-called
' cases of tapeworm' had been more or less misman-
aged. In fact, the entire series were ' cast-off' cases,
so to speak; and, indeed, I am not aware that my
opinion has ever been sought for in any instance
where the patient had not previously undergone'
treatment. At all events, I ascertained this to have
been the case in 78 out of the 80 patients here referred to. Only the presumedly difficult cases, or
such as are often called ' obstinate,' have hitherto
come under my observation. Of the more chronic
cases I will only say that one of the patients admitted that he had found it necessary to place himself under the care of twelve professional friends in
succession. As the entire profession is perfectly
well informed as to the therapeutical resources of
our art in this connection, it shows that immediate
success in the treatment of the disorder is not always
nor even usually dependent upon the choice of a
particular drug.

This latter observation naturally leads me, in
the next place, to say a few more words concerning

the results of previous treatment, so far as I could
gather from conversation or correspondence. Thus
it appears that out of 63 cases (in all of which it
was clearly ascertained that one worm only had
been present), there were no less than 40 instances
in which the entire body of the parasite had been
dislodged, and in 10 or 12 of these this result had
more than once been achieved. In a few cases the
body had been expelled by treatment on six or eight
separate occasions. In the remaining 23 cases
either small fragments of the strobile or merely
several loose proglottides had been discharged.

Amongst the remedies employed, male fern has
been the favourite. My notes on this point are not
complete; yet I ascertained that this popular drug
had certainly been administered in 28 instances,
turpentine in 13 instances, and kousso in 11 cases;
the other less commonly employed remedial agents
being pumpkin-seeds, kamala, tar-water, calomel,
aloes, sulphate of copper, santonine, &c. In particular and individual cases four or five of these agents
had been employed in succession. As regards my
own treatment of these self-same cases I find that in
the mere matter of drugs I have resorted to male
fern in 52 cases, and in 39 of these it was the only
remedy employed. In the 13 remaining cases either
areca-nut, kamala, or kousso was resorted to, frequently in combination with male fern in the simple
form of powder. Only in such cases as those

in which other parasites were either known or suspected to be present have I enjoined the use of santonine and certain other drugs, which I deem utterly
unsuitable as tæniafuges or tæniacides.

In regard to the special results obtained by
myself in connection with these 80 miscellaneous
'cases of tapeworm,' I have, in the first place, to
dismiss the 24 delusive ones, and to them I must
also add 9 others in which either a simple opinion or
merely preliminary advice was offered. This leaves
48 cases only in which I had any chance of testing
the action of appropriate tapeworm remedies. In
exactly 24 out of the 48 I ascertained that the
cure was complete, and, except in some six or eight
of the remaining patients who refused to persevere
in the treatment, I believe that the same good result followed. Only in a comparatively small proportion of the cases have I been enabled to make
the necessary, complete, and final stool inspections;
by the results of which alone could I be enabled
to pronounce a cure or otherwise on the dismissal or
departure of the patient. However, I now usually
insist upon this essential detail of management being
properly carried out. Thus, in 17 out of the 24
cures above mentioned, the head of the worm was
actually obtained, and in the 7 other cases it was
ascertained that the parasite never returned after
treatment.

The management of genuine cases of tapeworm

is incomplete without attempts on the part of the practitioner to secure the head by personal investigation of the matters discharged. It is not well to leave this task to others. On four separate occasions I have known the patient to remove the head of the worm with the rest of the parasite from the stools; but, with one exception, these persons were unaware of the character and importance of their individual ' finds.' Guided by my own more recent experiences, I think we ought always to reckon upon curing straight off four out of every five cases that are presented to us for treatment, provided, of course, we have the proper facilities offered to us. However, it is only in a small proportion of cases that the necessary examinations can be efficiently made. Thus, in 1873 I enjoyed this opportunity of searching for the head of the worm in six cases only, but in every one of these cases I removed the head of the tapeworm from the matters discharged. In every case also the head was found detached from the body of the worm, and in one or two instances it was completely isolated from the neck.

In connection with this brief analysis there are many other suggestive points. The relatively greater prevalence of the beef tapeworm as compared with the pork tapeworm in these cases bears out what I have elsewhere taught as the result of the examination of a much larger number of tapeworms sent to me by medical friends. I believe

that not more than two of the tapeworms expelled
in the above cases were examples of *tænia solium*;
and I reckon that, except amongst the poorer classes,
this last named species is not encountered in this
country in more than 5 per cent. of the cases occurring in general practice.

I venture to assert that the foregoing data, notwithstanding their fragmentary character, are amply
sufficient to prove that the diagnosis, prognosis, and
successful treatment of tapeworm is not such a
trifling matter as some persons seem to think. An
ordinary druggist's assistant can very well perform,
the function of prescribing and making up a male-
fern mixture; but the proper management of tapeworm cases is largely dependent upon an accurate
knowledge of the structure, habits, and general
economy of this singular class of parasites.

Altogether ignoring, or at least failing to recog-
nise, the practical value of these researches, it would
appear that the method of treating tapeworm on the
Continent—as expounded by the members of the
Society before referred to—is still made a matter
either of mere drugging or of a mere choice of drugs.
Thus one gives pumpkin-seeds, another pomegranate
root bark, a third male fern, a fourth kousso, and
so on, *ad nauseam* in more senses than one. It is
to be feared that a similar state of things prevails
at home. In England turpentine is still perhaps
the most popular remedy, at least in country dis-

tricts. Unquestionably all these particular remedial agents have their value, some being more con-
spicuously useful than others. As to the amount of
intellectual capacity requisite for the mere administration of either the one or the other, perhaps the
less said about that the better. Certainly, nothing
effective can be done without drugs, and, whatever
credit may be accorded to any medical practitioner
on the score of selection, the pharmaceutists ought,
in my judgment, to receive the first thanks. With
infinite care and trouble they have succeeded in
giving us some very choice and convenient preparations, and I do not think they have received that
share of the credit which is their due in this respect.
But, I repeat, the successful treatment of tapeworm
is not a mere matter of the choice of anthelmintics,
neither is it necessarily dependent upon the degree
of drugging. The pharmaceutists have executed
their part of the business, so to speak, almost fault-
lessly ; but the practitioner's functions have for the
most part been conducted in an incomplete manner,
and therefore inefficiently to a greater extent than
would otherwise have been the case. Without
doubt, it would be unjust to maintain that perfect
cures of tapeworm were rare; nevertheless, I have
before me absolute proof that, under the measures
commonly employed, these cures are neither so frequent nor so rapid as they ought to be.

In a work of this kind I cannot give very full

particulars of any of the cases without unduly increasing the size of the volume; nor can I comment upon them to the extent their separate importance demands. However, as illustrating some of the lessons to be deduced from their consideration, I will speak of the first case—abridged (at p. 53) from the fuller account which was published in the 'Lancet' for December 1874 (p. 795).

In the first place this case shows that success is not unattainable even in instances where the youngest children are afflicted with tapeworm. In the next place, it shows that there are examples of the disorder where it is better to proceed cautiously than to give the largest doses the patient can bear at once. Further it shows that the male fern may be administered every other day for weeks in succession without bringing away the *head* of the parasite, and that the employment of this powerful drug may give rise to alarm, if not on the part of the practitioner, at all events to the patient's friends. It likewise shows that personal investigation of the fæces affords (to anyone familiar with the facts of tapeworm development and appearances) a true indication of the proper extent to which the treatment has been pushed. Thus, whilst many practitioners would have been contented to have discharged the case when they had learned that the entire body of the worm, or 'several yards,' had been dislodged, I, on the other hand, felt sure that the almost perfectly

isolated head remained behind, and that a second
dose of the extract could be borne without ill results
to the patient. Had I been content to have let the
matter rest, the still attached head would undoubtedly have grown again, and some third person
might have been entrusted with the case. We
have seen, I repeat, that the very same drug, in
doses just as large as those I eventually administered, proved ineffectual though continued for a
lengthened period, the body of the parasite being
dislodged, but not the *head*. Now, supposing the
head had been expelled, since no special investigation was conducted with the view of ascertaining
that fact, it is obvious that it would not have been
known when the head actually made its escape;
and thus, as often happens, the treatment would
have been needlessly continued for a greater or less
length of time. Patients have come to me who
have been thus roughly handled for several years
after the head of the tapeworm had been dislodged.
In this case the treatment was not continued longer
than was absolutely necessary. Some may contend
that this accurate knowledge of the facts of these
cases, whether negative or positive, is of no importance. I beg very respectfully to differ from these
individuals, some of whom have been pleased to
favour me with probably the most discourteous ano-
nymous communications that any honest labourer
in the cause of practical science ever received.

At the risk of receiving further discourtesy from a few eccentric individuals, I do not hesitate again to say that, to go on persistently drugging in total ignorance as to whether the head of the tapeworm is present or not, is, in my humble opinion, a slovenly mode of procedure; and even in instances where the practitioner does happen to know that the head actually remains intact, it is sometimes advisable to defer anthelmintic treatment. In the present case, I was certainly particularly fortunate in securing the co-operation of an unusually intelligent nurse, and thus not a single loose proglottid escaped per anum without subsequently undergoing the neces-· sary scrutiny. But why scrutinise them at all? some unfriendly hypercritic will say. To persons possessing such ill-adjusted mental processes I know it is useless to make reply. There are others, however, who set a different value on the lessons to be gathered from experience. Let me say, therefore, that it is within my personal knowledge that scores of different foreign bodies have been mistaken by inexperienced persons, both professional and otherwise, either for portions of the body of the tapeworm or for the head itself. I am constantly receiving such productions from medical friends for identification or determination. All sorts of delusions afflict patients also in this respect; thus I have recently had under my care a gentleman who was persuaded he had seen the head of one worm which infested him, although the said head was described as being larger than the sexually mature proglottid itself! Other patients have repeatedly, during my examinations, flattered themselves that they could pick out the head of the worm, but, with one exception, they abandoned the search in sheer weariness and disgust at their fruitless attempts.

. The medical profession cannot justly accuse me of withholding any information which my labours in parasitology (extending over a period of more than a quarter of a century) have produced. Besides several separate works, I cannot have published less than 200 papers and memoirs on the Entozoa. Notwithstanding these sustained efforts, a single lecture of mine, published in the pages of the 'British Medical Journal,' formed the subject of unkind comments at the hands of a few of my medical brethren. I certainly had made important omissions, which I hastened to rectify. The express object of that lecture, as subsequently stated in my *addendum* (in the 'British Medical Journal' for Jan. 24, 1874), was to insist strongly on the necessity of persistent and careful personal search by the medical attendant of the stools of his patient, until he was satisfied that the neck and entire head were found. To some practitioners, this seems apparently a trivial and easy thing to do. On this score, however, I have reason to know that it is not easy, and that the duty is frequently, in many cases habitually, omitted; and hence we have one important cause of failure to cure tapeworm.

No portion of the fæces should be left uninvestigated. On many occasions I have remonstrated with patients and attendants, who, from motives of delicacy, have, prior to my visit for the express purpose of ascertaining results, carefully removed all floating matters of a non-fæcal character. I have obtained proof that the head of the worm is liable to adhere to such light materials.

Of course, in conducting these researches, one is apt to encounter rather unpleasant experiences. Some of these I recounted to my class with the view of impressing my hearers with the necessity of performing this duty in a thorough manner. The examination should extend to every stool while the patient is kept under treatment. As I invariably employ disinfectants, the disagreeableness of the task is not so marked as many might suppose; nevertheless I have sometimes been attacked with vertigo and nausea, consequent upon several hours' search in a stooping posture.

By way of illustrating the proceedings necessary in some cases coming within my experience, I narrated in the lecture, which originally appeared in an abbreviated form, the following case:—A practitioner of repute having repeatedly attempted and failed to cure his daughter, solicited my advice, by

E

letter, in the first instance. I prescribed male-fern extract in the ordinary way, and recommended a search for the head after the exhibition of the tæniafuge. As, however, at varying intervals, the worm continued to reappear, I at length advised him to send the patient up to town, which he did. I again prescribed the oil of male fern. The body of the worm came away, as usual. After removing from the stools some ten or eleven feet of the body, including many loose proglottides, I continued the search for about two hours, during which I obtained some fragments of the neck, which were so fine from the part near the head that the transverse lines of segmentation were barely discernible. Ordering a small dose of castor-oil, with plenty of weak tea and warm milk, I promised to renew the search next day. I felt confident the head must have been dislodged from the upper bowel. When I called next day, the fresh evacuations had been retained in accordance with my instructions, and the investigation was renewed. Greatly advantaged by the strong light coming through the large hotel window, it was not many minutes before I noticed in the fæces four minute dots of uniform size. As these mere specks were placed side by side in a symmetrical manner, I concluded that they were the suckers of the head of the tapeworm. The conjecture proved to be correct. Much, if not fully half, of the head was gone, whilst the loose and partially disconnected suckers only remained attached to one another by fine connecting shreds of parenchyma. On carefully transferring them without injury to a small glass tube containing water, the help derived from a powerful pocket lens at once set every trace of doubt at rest. My previous day's examination having afforded proof that only one tapeworm existed, I had now no hesitation in giving the necessary assurances as to the completion of the cure (see Case IV. in this work).

I must not omit to state that I transferred the corked tube to my waistcoat pocket, and when, after walking a short distance, I again looked at the specimens, the four suckers had separated. The gentle agitation produced by walking had sufficed to break up the delicate connecting filaments.

According to Professor F. Mosler, of Berlin, great advantages are to be obtained by the employment of large enemata in the treatment of tapeworm. He thinks that the head of the worm can refix itself in the colon after expulsion from the upper bowel. I believe this notion to be entirely at variance with fact; but when the head lies loose in the lower bowel, the exhibition of an injection will certainly facilitate its final expulsion, and circumstances may arise to render such a step advisable. A patient, for example, may be in such a weak state as to be unable to endure the administration of powerful cathartics, or, at all events, a succession of cathartics, the exhibition of which, in very obstinate cases, is sometimes necessary. In such instances only should the enemata be resorted to, and only after there is evidence that the entire strobile has been dislodged.

CASES.

THE clinical value of the following record is necessarily somewhat invalidated by the fragmentary character of the data offered. So variable are the experiences encountered in the proper management of cases of tapeworm that it is almost impossible to generalise on the subject. In some instances the symptoms are of the gravest nature, whilst in other cases there may not be even a suspicion of the parasite's presence on the part of the victim. One thing is certain. The bearer of tapeworm runs considerable risk, if not of his life, at all events of deranged health, which may become so serious as to be the cause of great anxiety to himself and friends. The cases here given, notwithstanding the meagreness of the details, abundantly illustrate this truth. They must in some measure be left to speak for themselves.

CASE I.—E. L. This was a little girl, only two years and three months of age. Her parents consulted me in the month of February 1874. She had contracted the disease about nine months previous to the date at which I saw her, and had been treated with considerable success, at least so far as the expulsion of the body of the worm was concerned. She had suffered severely in her general health; more from the drugs, I believe, than from the parasite itself. Fortunately, I obtained the *head* of the worm, though advising under circumstances of peculiar difficulty. The case is fully reported in the 'Lancet' for Dec. 5, 1874. Cured.

CASE II.—E. S. B. In December 1874, I saw this little boy of five years of age. He had been treated for nine months, the body of the worm having been secured in its entirety, but not the head. His symptoms, though not severe, gave much anxiety to his parents. I employed male fern, and obtained the *head* of the worm. In this case the drug had a slight toxic effect upon the patient, but the unpleasant symptoms passed off in less than half an hour. Cured.

CASE III.—A. A., March 1871. This was a female infant, only eighteen months old, who nevertheless had contracted tapeworm four months previously. I was merely requested by the ordinary medical attendant to give an opinion. In so young a child I considered that the further employment of male fern (although already given with considerable success as regards the worm) would be attended with risk, and consequently advised the employment of areca-nut powder in its stead. I do not know the ultimate result. The case is of interest as showing at how early an age the parasite may be contracted. Other particulars are given in my book on ' Worms,' p. 72.

CASE IV.—W. W., Dec. 1872. This was a little girl, three years of age, the daughter of a medical gentleman residing in Somersetshire. Her symptoms occasioned much alarm to her parents, who, after the lapse of a year, during which time she had been repeatedly treated with partial success, sent her to town. By employing male fern I ultimately dislodged and obtained the completely isolated *head*. The case is briefly recorded in the ' Lancet,' for Dec. 19, 1874, and also more fully in the ' British Medical Journal,' for Jan. 24, 1875, where a figure of the partially disintegrated head of the worm is given. I regard this case as one of the most instructive with which I have had to deal. Cured.

CASE V.—T., July 1872. This was the case of a little girl who contracted tapeworm in China. Her age at the time I saw her was three years and nine months, the worm having been observed about six months previously. By treatment, several yards of the parasite had been expelled, but not the *head*. In this instance I prescribed kamala, areca-nut and male fern in every convenient form that could be devised, but only portions of the medicines were retained by the stomach. The male fern produced ill effects, and the case was provisionally abandoned on account of the sensitive condition of the stomach.

CASE VI.—C. M., May 1869. This was a little girl of four years of age, the daughter of a distinguished medical officer. She had contracted tapeworm in India, and was brought over to this country for treatment. The symptoms were comparatively slight, but they caused anxiety to her parents. I had occasion to treat her several times before the *head* of the parasite yielded, which it did at a time when I was deprived of the opportunity of searching for it. Her case is more fully described in my volume on ' Worms,' p. 21. Cured.

CASE VII.—C. B., April 1871. This was a lad, six years of age, who for three years had played the *rôle* of host to the tapeworm guest. The symptoms, though not severe, were by no means slight, his general health being rather seriously disturbed. I had no opportunity of watching this case, but I prescribed areca-nut. The worm was ultimately discharged, but whether in consequence of the action of the drug I prescribed I could not clearly ascertain. At all events, there was no return of the tapeworm. See ' Worms,' p. 71. Cured.

CASE VIII.—W., September 1872. This was a lad, eleven years of age, who had contracted tapeworm eighteen months previously, his mother being also the victim of another parasite apparently about the same period. He had previously obtained relief at intervals from the usual medical attendant, who solicited my opinion in consultation. In this case the symptoms complained of were chiefly languor and pains in the back. I advised the employment of areca nut, but am unaware as to the result.

CASE IX.—H. A., October 1869. This was a young lady, thirteen years of age, who had contracted tapeworm only four months previously. She had, nevertheless, been several times treated by a druggist without success. Her symptoms were so severe that in the first instance I counselled delay as to treatment. At a subsequent period she came to town, when I treated her with areca-nut powder and male fern, eventually securing the *head* of the parasite. For particulars, see 'Worms,' p. 41; and also the 'Lancet' for December 19, 1874. Cured.

CASE X.—H. C., May 1870. This was a lad of thirteen years, who had played the part of host to a tapeworm for fully three years. He was suffering from chorea at the time I saw him, and as he had recently parted company with the body of his guest, I counselled delay in the matter of vermifuges, and prescribed tonics. The case is pretty fully recorded in my volume on 'Worms,' p. 64, where additional reasons are given for my withholding the active treatment which his parent almost demanded.

CASE XI.—H. J., July 1874. This was a lad,

fifteen years of age, whose case was worse than the previous one in respect of the nervous symptoms. The chorea, accompanied with debility and emaciation, was represented by his father as extreme. I did not see this patient, and gave general advice only by letter. The lad had been treated at a dispensary, large portions of the worm having been from time to time dislodged. The most remarkable thing is that this poor lad seems to have been the victim of tapeworm ever since he was four years of age. The case is further instructive as showing the necessity of early securing the *head* of the worm.

CASE XII.—S. A., January 1874. This was a lad of seventeen years of age, who had harboured a tapeworm for six years, and for which he had been repeatedly under treatment with more or less partial success. His symptoms were chiefly those of lassitude and vertigo, rendering his duties irksome. I dislodged all the body and neck of the worm, close up to the *head*, after which the treatment was discontinued.

CASE XIII.—L. R., October 1864. This was the case of a student, eighteen years of age, who, though harbouring a large tapeworm, was only made aware of the fact by listening to some observations that I had been making on the subject of worms. He employed the oil of turpentine as a remedy, and put himself in communication with me as to the results. This is the only instance in which I have ascertained with certainty that turpentine caused the expulsion of the *head* of the tapeworm. See 'Worms,' p. 24, and 'Lancet' for December 19, 1874. Cured.

CASE XIV.—S. G., July 1865. This was the case also of a young man, eighteen years old, who, however, had been conscious of the presence of his enemy for a period of five years, during which his sufferings were of a very grave character. As in the above case, my opinion as to the results of treatment was solicited, and from the materials sent me in a bottle I had the good fortune to find the *head* of the worm. See 'Lancet' for December 19, 1874, and also 'Worms,' p. 26. Cured.

CASE XV.—S. S., October 1873. This was likewise an instance where the bearer of the tapeworm was a young man, eighteen years of age, who, however, had only enjoyed the privilege of playing the *rôle* of host for a single year. His symptoms were not remarkably severe, and he had been treated with partial success on several occasions. I was only called upon to give preliminary advice, although the case was one that was eminently favourable for active treatment.

CASE XVI.—G. G. In this instance also, that of another lad, eighteen years of age, my opinion only was solicited. The patient had suffered rather severely for two years past, the nervous symptoms being ascribed either to tapeworm or to some other internal parasite. At all events he had been treated for worms by a dozen medical men without experiencing any other than a negative result. As regards tapeworm, I satisfied myself that this patient and his friends were labouring under a delusion, and advised accordingly.

CASE XVII.—W. L., June 1870. This lad, nineteen years of age, was the son of a butcher. He had, I believe, contracted the worm about a year previously to the time I saw him; and, as the result of treatment, had very shortly before passed, it was said, as many as thirty feet of tapeworm. I consented to prescribe a vermifuge. The worm did not, I understood, subsequently return; but whether the head was expelled by my remedy or not, it is impossible to say. At all events he was cured. See ' Worms,' p. 60.

CASE XVIII.—J. B. This young gentleman, twenty-one years of age, is a clerk whose business carries him much abroad, and he believes, with good reason, that he contracted tapeworm in Holland, about six months previous to the time I saw him, in consultation with his ordinary medical adviser. This was on January 11, 1875. He had a voracious appetite, and complained of a haziness or dulness of vision. His fondness for underdone meat was marked. I recommended the employment of male fern, and prescribed accordingly.

CASE XIX.—C. R. This was also the case of a young gentleman, a clerk, eighteen years of age.

His father consulted me, September 22, 1874, the son's symptoms being chiefly those of malaise and insomnia. His usual medical attendant had employed santonine, and had assured him that there was no tapeworm present. I expelled twenty-three feet of tapeworm, and also the *head*. See the 'Lancet' for December 19, 1875. Cured.

CASE XX.—M. W. This patient was also a clerk, twenty-two years of age, who had harboured tapeworm for three months only, but whose sufferings nevertheless were of the severest character. He had been treated by two druggists—one giving him santonine, and the other male fern. The case (November 24, 1875) was urgent, and although the parasite was not fully developed, I obtained not only the body but also the *head*. The case is more fully reported in the 'Lancet' for December 19, 1874. Cured.

CASE XXI.—B. C. This young lady, eighteen years of age, first consulted me in June 1874. She had harboured tapeworm for three months, and had been repeatedly treated with only very slight results. Having expelled the entire body of the worm, 23 feet in length, I counselled no delay. She went abroad and was treated there with an apparently similar result. The *head* was not found. After a long interval I was again consulted about her case, but the worm not having reappeared, I pronounced her cured.

CASE XXII.—C. F. This lady is the wife of
a city clerk, who consulted me in December 1866,
and is about twenty-five years of age. For a year
she had undergone fruitless attempts at treatment.
Her symptoms gave great annoyance to her husband.
Fortunately, in this instance, the employment of
the male-fern extract was speedily successful, dislodging the entire body of the parasite, and also
separately the head and neck. See the 'Lancet'
for December 19, 1874, and also 'Worms,' p. 35.
Cured.

CASE XXIII.—E. R. This gentleman had
entertained his tapeworm guest for more than sixteen years. I saw him in July 1873. The host
and guest seemed to be on such friendly terms that
the patient only occasionally thought it worth his
while to submit himself to medical treatment. He
had, however, several times taken large doses of
male fern. I employed the same remedy in small
doses, and expelled not only the body but also the
head of the worm. See the 'Lancet' for December
19, 1874. Cured.

CASE XXIV.—C. B. This was a man of
also about twenty-five years of age. He was the
patient of a City physician, who on two occasions
requested my opinion in reference to the prognosis.
The patient's second visit to me was made in June
1865, when I detected the *head* of the worm. The
case is the first recorded in my Lectures on Helmin-

thology, and is very instructive in reference to the growth of the parasite. See 'Worms,' p. 5. Cured.

CASE XXV.—B. E. This gentleman, a banker, had for the last two years suffered very severely from tapeworm. I saw him in June 1873. He had locomotor ataxy of the upper limbs, with obscure abdominal pains and other grave symptoms. He staggered about, and was frequently sick. The symptoms were aggravated by the presence of oxyurides. I expelled the tapeworm, and also its *head*, which was nearly isolated. See the 'Lancet' for December 19, 1874. Cured.

CASE XXVI.—H. J. This lady consulted me in November 1866. Her case was one of the most distressing character, since, although she had long parted with her tapeworm guest, she could not be persuaded that such a separation of host and guest had really been effected. The nervous symptoms were very severe. For further particulars, see my volume on 'Worms,' p. 12. Delusion case.

CASE XXVII. P. J. This patient, a clergyman, suffered rather severely from a tapeworm, which he had contracted about a year previously. The performance of his public and private duties was thus seriously interfered with. He consulted me in January 1873. Fortunately, a brief course of treatment sufficed to bring away the body of the worm, and also the *head* separately. See the 'Lancet' for December 19, 1874. Cured.

CASE XXVIII.—G. D. This young officer in the merchant service, aged twenty-two, consulted me in April 1875. He found he had tapeworm whilst cruising up the Persian Gulf. The employment of kousso and other remedies in large doses had had the effect of bringing away portions of the body, but from the nervous symptoms that remained he felt sure that he was still the victim of the worm. He had lately passed the entire body, but was not sure about the *head*. I enjoined rest, and discountenanced the further employment of vermifuges for the present.

CASE XXIX.—T. W. This gentleman, in active business, had entertained his guest for three years, despite frequent treatment by chemists. He had also played the *rôle* of host to lumbrici and oxyurides. I saw him in March 1873. Like the above-mentioned patient, he took a desponding view of the possibilities of recovery. I employed male fern, and expelled the entire tapeworm, including the *head* and neck. See the 'Lancet' for December 19, 1874. Cured.

CASE XXX.—G. W. G. This patient came from the United States desiring to be treated for tapeworm. I saw him on July 13, 1869. He had been heavily drugged with turpentine, having undergone treatment, off and on, for a space of six years. I sought to persuade him that he and his guest had already parted company, and I eventually proved to him that such was the case. See ' Worms,' p. 16. Delusion case. Cured.

CASE XXXI.—W. W. This gentleman contracted the disease in India. I saw him in May 1873. He had, during the last three years, been continually treated without success, and had become highly nervous and desponding. He did not believe that there was any possibility of his being cured ' after what he had gone through.' He was on his way home to the United States. I expelled the worm, and found its *head* on the following day. See the ' Lancet ' for December 19, 1874. Cured.

CASE XXXII.—C. A. In this case, that of a Scottish gentleman, the parasite (in reference to which I was consulted by letter in the month of June 1868) had evidently taken its departure for nearly a year; nevertheless the patient desired to be treated for its expulsion. As the means to attain a legitimate end, I prescribed, and convinced the patient that he was all right. See ' Worms,' p. 16. Delusion case. Cured.

CASE XXXIII.—W. J. This man had been troubled with and treated for tapeworm for more than two years. I was requested to prescribe for him in June 1873. I did so ; but left the general management of the case to others. The male fern

F

was employed at intervals, and eventually with success. I was afterwards told that the *head* was obtained. See the 'Lancet' for December 19, 1874. Cured.

Case XXXIV.—R. A. In the month of September 1869, this gentleman consulted me in reference to the cause of certain obscure pains in the throat and other parts of the body. He satisfied me that he had contracted tapeworm some years previously, but I was sure that host and guest had parted company. It was a case of genuine hypochondriasis, and I advised accordingly. Delusion case. Cured.

Case XXXV.—S. S. This was a lady whose symptoms were of a severe order, there being partial hemiplegia and cerebral irritation, with spectral illusions. I saw her in June 1867. She had involuntary spasms of the muscles of the left cheek. In a few days I expelled the tapeworm almost entire, the *head* being found separately detached. Her nervous symptoms disappeared. See the 'Lancet' for December 19, 1874; and also my volume on 'Worms,' p. 35. Cured.

Case XXXVI.—R. J. J. In this instance a period of five months had elapsed since the patient last enjoyed ocular proof of the existence of his tapeworm guest. He consulted me in the month of November 1867. Fortunately, after adopting some remedial measures, I persuaded the patient that his

enemy had departed. See ' Worms,' p. 17. Delusion case. Cured.

CASE XXXVII.—J. H. This gentleman had
undergone homœopathic treatment for tapeworm—it
is needless to say with what result. I saw him in
January 1867. His symptoms were fortunately
slight. For several days in succession the worm
resisted the action of considerable allopathic doses
of male fern; but I eventually dislodged both the
body and the *head* of the parasite, securing the latter
only after many hours of fatiguing investigation.
The case is recorded at length in ' Worms,' p. 36.
(H. H. J.). See also the ' Lancet' for December
19, 1874. Cured.

CASE XXXVIII.—H. H. W. This Indian
officer consulted me in September 1869. He had
contracted tapeworm in the Punjab some three or
four years previously, and still believed himself to
be the victim of the parasite. In that opinion his
usual medical adviser shared. I was satisfied that
he and his guest had long parted company, but I
did not so persuade him until I had ordered the
usual remedies. See ' Worms,' p. 17. Delusion
case. Cured.

CASE XXXIX.—C. B. J. This retired Indian
officer, thirty-eight years of age, had partial reflex
paraplegia, frequent vertigo, fits of insensibility, and
severe mental distress. He first observed the worm
five years previously, and had undergone, he said,

active treatment at the hands of 'scores of doctors.'
I saw him in August 1874, and employed the male
fern. The bodies of two worms were expelled with
their necks intact. I counselled delay in this case as
to further treatment—partly on account of the ex-
cited state of the patient. The parasites returned,
and I again expelled them. Both of the *heads*
were this time found by me, detached from their
respective necks. The worst nervous symptoms
at once disappeared. See the 'Lancet' for December 19, 1874. Cured.

CASE XL.—W. S. This gentleman, an en-
gineer from India, like the officer whose case is
given above, played the *rôle* of host to two tapeworms. In April 1867, I employed the male-fern
treatment, dislodging ten feet of tapeworm and also
one *head*. This patient did not afford me an opportunity of renewing my treatment in view of securing
the perfect expulsion of the second worm, preferring
to manage himself by aid of the prescriptions and
instructions I had previously supplied. I cannot be
sure that he got rid of both enemies, but of the first
he was certainly cured. See the 'Lancet' for
December 1874, and also my lectures on 'Worms,'
p. 27 (S. S. W.).

CASE XLI.—B. V. This gentleman, an Indian
officer, was long troubled with tapeworm, and his
case proved one of the most obstinate with which I
have had to deal. I several times, at intervals, ex-

pelled the body, and eventually the entire young tapeworm, including the *head*, before it had on this occasion arrived at sexual maturity. The case was complicated with threadworms, and the symptoms, to use the patient's own phrase, showed that the worms were ' playing havoc with his constitution.' See the ' Lancet,' as above ; and for further particulars the volume on ' Worms,' p. 68 (B. V. H.). Cured.

CASE XLII.—P. P. This was the case of a delicately nurtured lady, who, however, was living in reduced circumstances. She had been treated unsuccessfully with turpentine. From the strong representations made to me I was induced to prescribe for her. By the employment of male fern I expelled the entire parasite, which never afterwards returned. See ' Worms,' p. 29. Cured.

CASE XLIII.—M. M. R. This Indian officer, a captain in the Royal Artillery, had suffered from tapeworm for six years. He saw me in January 1866. The symptoms were rather severe, and his life was jeopardised by the fact that at times he experienced much difficulty in keeping the saddle. I expelled the worm, and some time afterwards understood that there had been no return of the parasite. See ' Worms,' p. 30, where further particulars are recorded. Probably cured.

CASE XLIV.—S. A. This gentleman had been long resident in Peru, S.A. He consulted me in

February 1869, complaining chiefly of pain in the region of the stomach. The tapeworm was probably contracted in England. My treatment not only expelled an ordinary tapeworm but also a solitary whipworm. The case is more fully recorded in my lectures on ' Worms,' p. 31.

Case XLV.—L. A. J. The husband of this lady, residing in Kent, consulted me respecting her in May 1869. The case proved a very instructive one, inasmuch as the *head* of the tapeworm was passed by the only evacuation I had not an opportunity of examining. The symptoms were severe, and the loss of health considerable. It was a very obstinate but successful case. See ' Worms,' p. 31. Cured.

Case XLVI.—Y. Y. This gentleman was an Irishman, who at the time of consulting me by letter (June 1866), had been suffering from a tapeworm for a period of about two years. I had no opportunity of seeing him ; but I consented to prescribe for him, and had the good fortune to expel the parasite, which never afterwards returned. See ' Worms,' p. 32, where the case is fully recorded. Cured.

Case XLVII.—B. C. C. This was the case of a lady who contracted tapeworm at Poonah, in India, some six months previous to the time of my seeing her, which was in the month of January 1868. Her husband, an officer in the Indian army, had also I believe contracted tapeworm about the same period. In this case the *head* of the parasite was dislodged separately from the body of the worm. See 'Worms,' p. 32. Cured.

CASE XLVIII.—D. B. This Jewish gentleman consulted me in December 1865. He had fainting fits and considerable mental depression—these and other unpleasant symptoms being, in his own judgment, entirely due to the presence of tapeworm. He had been under treatment for more than a year, without experiencing other than negative results. I employed the male-fern extract, bringing away the whole body of the worm, including the *head*. For further particulars see my lectures on 'Worms,' p. 34 ; also the 'Lancet' for December 19, 1874. Cured.

CASE XLIX.—P. H. This case came before me in November 1867, and is of interest on account of the circumstance that the patient contracted her tapeworm guest whilst resident in Germany, and also because there is every reason to believe that it was an example of the broad tapeworm or *Bothriocephalus latus.* Unfortunately the lady merely desired my professional opinion, not wishing to submit herself to the necessary treatment for the expulsion of the parasite. See 'Worms,' p. 58.

CASE L.—G. A. This lady, the wife of an Indian officer, consulted me in March 1871. Her case is certainly one of the most remarkable that has ever come before me. She stated that not only was she the victim of tapeworm, but she had been undergoing active treatment for the expulsion of the worm for a period of six years. She very rudely resisted the arguments I adduced to prove to her that she never had a tapeworm, setting up the opinions of my medical brethren against me. They were wrong, as I ultimately succeeded in convincing her. See ' Worms,' p. 44 to p. 49 inclusive. Delusion case. Cured.

CASE LI.—J. T. This patient's case is of interest as showing the utility of a change of remedies. He chiefly complained of pricking sensations within the abdomen. He consulted me in December 1870, but I had no opportunity of searching for the head of the parasite. In this instance I found kousso effective in bringing away the entire body of the worm, after scammony, areca-nut, and male fern had each been employed with very little success. See ' Worms,' p. 50, where further particulars are given.

CASE LII.—D. W. This patient, a resident in the east of the metropolis, consulted me in September 1870. The tapeworm resisted the action of considerable doses both of male fern and areca-nut, small portions only of the body of the parasite coming away. It seemed to me a suitable case to try the action of kousso, especially as I should have an opportunity of personally ascertaining results. However, he refused further treatment, despairing of success. For further particulars see ' Worms,' p. 52.

CASE LIII.—E. C. This lady came from New York, U.S., and consulted me in June 1870. In the short space of six months her general health had suffered considerably, so that she had, as she said, rapidly lost flesh. The exhibition of male fern, followed by a cathartic, brought away first eight and then nine feet of the parasite. The treatment was not pushed further, and I saw no more of her. See ' Worms,' p. 57, for further details.

CASE LIV.—L. D. This patient, a well educated superior servant and companion, consulted me in May 1870. She had harboured tapeworm for a period of three years, for which she had already taken most of the vermifuges commonly employed, with feeble results. On two or three occasions I removed the entire body of the parasite, male-fern and areca-nut being the remedies exhibited. For further particulars see ' Worms,' p. 58.

CASE LV.—C. M. This gentleman sought my advice in the month of April 1869. His symptoms were rather severe, debility and exhaustion being marked. He was not actually passing joints. I counselled delay until the worm gave ocular evidence of its existence. I then prescribed for him, and had the satisfaction of dislodging a tapeworm sixteen feet in length. See ' Worms,' p. 61, where full particulars are recorded.

CASE LVI.—S. L. T. This gentleman, an
officer in Her Majesty's Indian army, called upon
me in the spring of 1870. He had suffered much
from jungle fever, and was under the impression
that he was still the victim of a tapeworm for which
he had been treated some three years previously. I
perceived it was otherwise, but at his request prescribed an anthelmintic. The result showed that
my opinion was correct. See ' Worms,' p. 66. Delusion case. Cured.

CASE LVII.—R. M. This patient, a gentleman, sought my advice in May 1871. He was extremely agitated, and had evidently suffered severely
from the effects of a tapeworm, which, however,
I judged to have quitted its residence, as a consequence of treatment, head and all. My opinion
here also proved correct, and his general health
improved under the advice I gave him. See
' Worms,' p. 66. Delusion case. Cured.

CASE LVIII.—A. A. This gentleman's case
was in several respects a singular one. He came
from Liverpool. When I saw him, in October 1870,
he had been the victim of tapeworms for about four
years, his nervous system having suffered severely,
as was proved by the locomotor ataxy of the lower
limbs, and twitchings of various muscles. I advised
delay as to active treatment, as it was more than
probable that he and his guests had parted company. He appeared to have harboured the cucu-

merine tapeworm. See 'Worms,' p. 74, where details are offered.

CASE LIX.—A. H. This lady also came from Liverpool, placing herself under my care in the month of September 1871. Like the male patient above mentioned, she had probably been the victim of several tapeworms at once, and had, it was alleged, passed seventy feet of worm at a time. She had one parasite only when she consulted me. I expelled the entire body to the extent of twenty-four feet. See 'Worms,' p. 78, for further particulars.

CASE LX.—H. M. This was one of those characteristic instances where the patient erroneously attributed her severe hysterical symptoms to the presence of tapeworm. This young lady consulted me in October 1869, and her case was one of the most pitiable I have ever encountered. She was anæmic to the last degree, and had both diarrhœa and leucorrhœa. For full particulars, see 'Worms,' p. 74. Delusion case.

CASE LXI.—K. H. This officer in Her Majesty's Indian army consulted me in the month of July 1869. He had contracted tapeworm four years previously, and suffered considerably from nausea, loss of appetite, and general debility. Treatment had been pursued previously with very partial success. By the employment of male fern I succeeded in expelling the body of a pork tapeworm, sixteen feet in length, but no further efforts were made in view of completing the cure.   See ' Worms,' p. 20.

Case LXII.—G. J. M.   This gentleman, an M.P., probably contracted tapeworm in Norway, and, up to the time that I first saw him, he had already been victimised for a period of at least ten years.   He consulted me in September 1870.   On four separate occasions I expelled the entire body, as well as the greater part of the neck of the parasite, obtaining, in all, upwards of seventy feet.   The patient declined further treatment.   For full particulars, see ' Worms,' p. 52.

Case LXIII.—S. W.   I was requested to prescribe for this female, a servant, twenty-eight years of age, in the month of November 1871.   She was anæmic, and complained of faintness and vertigo. On two occasions I expelled the body of a worm, each about fourteen feet in length.   They were associated with threadworms, many of which were also expelled by the powerful action of the male fern.   The parasite did not return after the second course of treatment.   Cured.

Case LXIV.—J. S.   This patient, residing in the neighbourhood of London, consulted me in the month of January 1872.   He had contracted the parasite some eight months previously, and had been treated with some success; at least, he brought me several feet of tapeworm for inspection and identification.   His symptoms were severe, as he had partial locomotor ataxy. I prescribed for him, and expelled the entire body of the worm, but I had no opportunity of searching for the head.

CASE LXV.—J. G. S. This was the case of a young merchant resident in Spain. His father consulted me in the month of May 1872. I had no opportunity of seeing the son, but at the earnest request of the parent I consented to prescribe for him. It seemed from the father's statements that the young man had been recommended by a Carlist officer to put a certain remedy into a basin or stool, over which he was to seat himself until the drug brought away the parasite. Whether my prescriptions had more effect than the 'charm' I have not been informed.

CASE LXVI.—N. B. F. This gentleman, a New York merchant, consulted me in the month of May 1872. He had been troubled with tapeworm for nine months, and had taken pumpkin seeds, turpentine, and male fern, without any success whatever. The last-named remedy, in my hands, succeeded in dislodging the entire body of the worm, including the neck, to the extent of sixteen feet. No further treatment was employed.

CASE LXVII.—S. H. This individual, the private secretary of a distinguished personage, consulted me in the month of July 1872. He had apparently contracted the tapeworm in Turkey, and had been treated with partial success on five or six separate occasions; that is to say, portions of the body of the worm had frequently come away. By the employment of the male fern I expelled the entire body and neck of a tapeworm eighteen feet in length; but I never could gather whether the parasite reappeared, though I frequently solicited information on the subject.

CASE LXVIII.—M. N. This unmarried French lady consulted me in the month of August 1872. It was one of those distressing cases where the patient erroneously fancied herself to be the victim of worms. There were the usual gnawing pains in the region of the stomach so frequent in hypochondriacs. The movements of the parasite were also distinctly felt. I satisfied myself that she had parted company with her tapeworm guest some twenty years previously, and advised accordingly. She had also passed a lumbricoid worm two years since. Delusion case. Cured.

CASE LXIX.—A. W. This was another instance of painful error as to the source of severe symptoms which were attributed to the presence of one or more worms. This gentleman consulted me also in the month of August 1872. His conversation was rambling throughout, and he entertained all sorts of strange notions. He gave the most incoherent utterance to his opinions, at the same time candidly stating, however, that he felt it quite unsafe for him to be left alone. I advised an immediate change of scene, which was happily attended with beneficial results. Delusion case.

Case LXX.—B. R. This case was somewhat similar to the above, but the symptoms of nervous excitement were unaccompanied with any very marked delusions, save such as referred to belief in the existence of a parasite as the cause of the disorder. The patient, a gentleman in business, consulted me in October 1872. He had been overworked, and suffered from hemicrania, vertigo, and exhaustion. I prescribed belladonna and ammonia, likewise recommending an immediate change of scene. Delusion case.

Case LXXI.—R. R. T. This case bore a considerable resemblance to the preceding, the patient firmly believing that he was the victim of a tapeworm, of which he had evidently been cured long previously. The hypochondriasis was sufficiently marked when I saw him, and this was in March 1873. The case appeared to me to be one of incipient dipsomania, and as there was a free deposit of lithates in the urine, with much acidity, I recommended the Vichy mineral waters. Delusion case. Cured.

Case LXXII.—B. C. F. This gentleman originally visited me in February 1874. He was satisfied that he was still infested by a tapeworm, which he had contracted some time previously in Ceylon. He had the usual gnawing pains of hypochondriac patients, and therefore felt perfectly sure that the worm was still in his interior. I prescribed a vermifuge by way of testing for its presence, and I fortunately succeeded in persuading my patient that he and his worm had parted company for ever. Delusion case. Cured.

CASE LXXIII.—J. T. Of a similar order to the foregoing is the present case. This clerk consulted me in the month of December 1873. He contracted tapeworm some nine months previously, and had been treated for it with vigour. I counselled delay in the first instance, especially as the body of the worm had been expelled. He had taken copper, aloes, turpentine, and the pomegranate root bark. I subsequently ordered male fern, and satisfied both the patient and myself that he no longer entertained the tapeworm guest. Delusion case. Cured.

CASE LXXIV.—A. T. This was likewise an instance of suspected tapeworm, but as only ten weeks had elapsed since the body of the worm had passed as the result of previous treatment, I counselled delay, and suspended my judgment. In the event of the return of the parasite I was to be informed, with a view to curative treatment. He originally consulted me in the month of May 1873, when I prescribed a tonic and gave other advice in harmony with the data above mentioned. I received no further communication. Probable delusion case, following a cure.

Case LXXV.—D. M. The wife of a medical gentleman, whose husband requested my advice in May 1873. Her symptoms were severe, and she had been treated with only very partial results. I prescribed male fern and areca-nut in combination, and was afterwards informed that the body of the worm had been expelled. She also subsequently passed a large round worm, for which, in case others should be present, I recommended santonine and scammony. This patient's general health, however, was fatally affected from other causes.

Case LXXVI.—B. A. In the month of December 1874, I was consulted by letter respecting the case of a young lad, eight years of age, residing in one of the Channel Islands. He had been the victim of tapeworm for nine months, and during that period had thrice passed the body of the worm as the result of treatment. A West End firm of druggists had recommended the parent to consult an eminent operating surgeon, who has since had the courtesy to inform me that he knows as much about tapeworm as I do. Knowing the necessity of personal superintendence in view of effecting a cure, I considered this a case in which the child should be brought to town, and advised accordingly, but I heard no more from the ' anxious ' parent.

Case LXXVII.—A. C. C. The case of this little boy, two years and four months old, from Ireland, was brought under my notice on March 17,

1875. For more than a year he suffered severely
from convulsive fits, and had become much emaciated. He had passed a round worm; and had been
treated with male fern, turpentine, and santonine.
The mere sight of medicine was sufficient to produce
a ' fit.' Under these circumstances I had to proceed with great caution; first of all employing
kamala, and afterwards the extract of male fern.
On March 24 I dislodged the entire body and neck
of the parasite, seven feet in length. After this the
child's general health rapidly improved, but the
worm returned in July, when I again expelled it.

CASE LXXVIII.—M. W. I was called in
consultation respecting this lady's case in September
1872. She had suffered from tapeworm for about
two months, whilst one of her sons had been victimised for about a year and a half (Case VIII.). She
was highly nervous, complaining of sharp pains in
the head and obscure sensations in various parts of
the body. I prescribed the ethereal extract of male
fern, and left the subsequent management of the
case in the hands of her ordinary medical attendant.

CASE LXXIX.—A. L. This clergyman had
been long under the advice of practitioners of dis-
tinction, but I doubted if the alleged parasites for
which he had been treated were joints of tapeworm. He consulted me in March 1869. The
case was clearly one of hypochondriasis, and accordingly I advised a change of scene, from which he

derived a temporary benefit. See 'Worms,' p. 15.
Delusion case.

Case LXXX.—D. II. In the month of May
1874, I was requested to treat a little girl, four and
a half years of age, for tapeworm, from which she
had suffered for about six months. Small doses of
kamala and areca-nut producing no result, I ordered half a drachm of the male-fern extract. No
proglottid appearing, I counselled delay, doubting
if the worm were present. As she had passed several
yards only three weeks previously, there could only
have been a small portion left. The parasites
returned, and she was again placed under my care
in April 1875. In a few days I expelled about
eighteen feet of tapeworm, but it was rendered un-
certain if the head came away.

Case LXXXI.—N. N. In the month of
March 1874, a medical practitioner requested my
opinion in reference to some foreign bodies that had
been passed *per anum* by a lady who was being
treated for tapeworm. The foreign matters consisted
principally of vegetable *débris*, whilst certain coils of
detached mucous membrane several inches in length
bore only a very slight resemblance to worms. The
treatment recommended was, therefore, principally
directed to the dyspeptic symptoms which were
complained of. Delusion case.

Case LXXXII.—P. R. This gentleman, a
merchant, sought my advice in March 1874 ; but as

he could not stay to undergo the necessary course of treatment, I was limited to a consultation. The case was interesting, inasmuch as the patient had contracted the disease in France many years previously, and had fruitlessly subjected himself to treatment by means of kousso and turpentine. Though a fine-built man his timidity was extreme. He feared lest the action of the proposed remedies should produce serious personal discomfort.

CASE LXXXIII.—W. F. This young lad, eleven years of age, was supposed by his father, a medical gentleman, to be suffering from tapeworm; and certainly there were symptoms such as might suggest the presence of parasites of one kind or another. I advised a short course of treatment with a view of testing for parasites, when it turned out that the patient's symptoms were due to the presence of lumps of raw carrot, of which he had secretly partaken. Delusion case.

CASE LXXXIV.—A. H. In this instance a medical practitioner solicited my advice in reference to the case of a domestic servant who had been the victim of a tapeworm for a period of two years. At my suggestion he employed the extract of male fern, which brought away the body of the parasite. When, some months subsequently, I inspected portions of the parasite, which had been preserved, I also ascertained that there had been no return of the worm. Cured.

CASE LXXXV.—D. R. This gentleman, a lawyer, consulted me in May 1874. He also had contracted tapeworm about two years previously, for the expulsion of which aloes and turpentine had been prescribed. Before taking the medicine, however, he asked my advice, when I recommended male fern as being preferable. He would not, however, give me an opportunity of making the necessary examinations. The entire body of the worm was expelled.

CASE LXXXVI.—C. C. An officer in H. M. Indian army had harboured tapeworm for a period of sixteen years, during which time he had frequently been treated at intervals. He contracted the parasite at the siege of Lucknow. I saw him in May 1874, and after a few days' treatment I expelled the entire worm. In this instance the neck of the worm came away first, the body being subsequently expelled under circumstances which precluded my searching for the head.

CASE LXXXVII.—H. P. In the month of June 1874, I was requested by a medical practitioner to advise and prescribe, if necessary, for a patient of his, connected with the post-office. This man had been the victim of tapeworm for several years, and when he came to me he brought with him the entire body of a fine beef tapeworm that had been expelled only the previous day. I advised delay. He called about three months afterwards to say that the parasite had never returned. Cured.

Case LXXXVIII.—W. M. This patient, an hotel proprietor's wife, consulted me in the month of May 1874. She had been treated for tapeworm for a period of seven years, and on one occasion had passed several yards of the parasite. From her gnawing pains and sinking feeling she was sure that she was still infested. I expressed my belief that she and her guest had permanently parted company. After a short course of treatment this view proved to be correct. Delusion case. Cured.

Case LXXXIX.—P. H. In the month of June 1874, a medical gentleman sent a patient to town requesting my opinion on a supposed case of tapeworm. The patient had been treated for about a year, during which time it was thought numerous segments of the worm had passed. However, I satisfied myself that the proglottides in question were merely vegetable products, and advised accordingly. Delusion case. Cured.

Case XC.—H. E. A gentleman in business, who likewise sought my advice in the month of June 1874, believed himself to be suffering either from tapeworm or some other kind of parasite. After showing him various examples of entozoa, and after gathering particulars of his case as to symptoms and the results hitherto obtained by treatment, I here also satisfied myself that no worm of any kind was present, and advised accordingly. Delusion case. Cured.

CASE XCI.—M. A. M. In the month of June 1874, I was also requested by a medical gentleman to prescribe for a lady in poor circumstances. She had been afflicted with tapeworm, it was supposed, for about seven months, but neither turpentine nor kousso had effected any other than a negative result. I counselled delay, not being satisfied as to the evidence of the existence of any parasite. At a subsequent period, to satisfy the patient's mind, I prescribed male fern as a test, but was not informed as to the result, which, I have no doubt, was negative. Delusion case.

CASE XCII.—B. V. This foreign physician, resident in London, consulted me in April 1875. His highly nervous and irritable state was partly attributed to the presence of a tapeworm which he had contracted about a year previously, and for which he had taken kousso and pomegranate repeatedly. On one occasion he had parted with several yards of the worm. Having myself tried an extract of kousso without result, I resorted to male fern, which detached the entire body and neck, eighteen feet in length. The head not having been found, its expulsion was rendered doubtful.

CASE XCIII.—D. G. This gentleman, also in business, had contracted tapeworm in Germany. When I saw him in the month of September 1874, he had undergone no treatment of any kind, and had only first observed the parasite some three months previously. Beyond slight sickness, there were no marked symptoms or other inconveniences. Promising to return for active treatment at a more convenient season, he left for the Continent, evidently on easy terms with his parasite guest.

CASE XCIV.—B. E., a lady teacher of music, first consulted me by letter in January 1875, and shortly afterwards in person. Rather over thirty years of age, she had been the victim of tapeworm for three years, having been treated by a herbalist and also by a practitioner (who employed kousso) with partial success. From a 'sinking feeling' and other nervous symptoms, she was sure the parasite was still present, though no proglottides had passed for six weeks. I discountenanced immediate active treatment, and she never returned, probably because her guest had departed.

CASE XCV.—B. C. E., an officer in the Royal Artillery, consulted me in the month of August 1874. He had severe paralysis from tapeworm, contracted in India, for fully a year and a half. He had taken kousso, kamala, turpentine, and male fern with partial success. When I saw him the paraplegia was so marked that locomotion was accomplished only with difficulty. I believe I expelled the entire parasite from this gentleman, but, through the carelessness of a servant, I was prevented from fully ascertaining the results of a very active course of treatment. The paralytic symptoms almost entirely disappeared.

CASE XCVI.—L. R., a musician, consulted me in November 1874. His symptoms were not attributed to tapeworm by his usual medical adviser, and it subsequently became clear that the obscure abdominal pains, muscular twitchings, grinding of the teeth, occasional wandering, and great irritability, were due in part to other causes than tapeworm. However, I caused the expulsion of a beef tapeworm, sixteen feet in length, and then counselled a. temporary change for the sake of the general health. The parasite returned, but it was not deemed prudent to renew the treatment, as he was suffering from uræmia, which at length proved fatal.

CASE XCVII.—P. P. C. A medical gentleman requested my advice by letter respecting the case of a lady who had been suffering from tapeworm for about a year. Thus far the treatment had been successful only to the extent of expelling numerous proglottides, and amongst those sent me for determination there was also the larva of an insect, the head of which had been surmised to be that of a tapeworm. Similar fragments of spurious worms have frequently been mistaken for tapeworm heads, and have been sent to me as such.

CASE XCVIII.—B. H. This is the case of an officer in Her Majesty's Indian army, who has been and probably still is the victim of a worm contracted seven years previously. He first consulted me in June 1874, when I expelled twenty-five feet of tapeworm, and again in the following September, when I obtained another twenty-one feet. In this instance I instituted a vigorous search for the head without success, but on one occasion I detected several larvæ of Anthomyia. As in three or four other cases, previously recorded, a third and possibly a fourth course of treatment may be necessary to effect a permanent cure.

CASE XCIX.—N. A. This gentleman's case, that of a merchant, thirty years of age, resembles the above in several particulars. He had contracted the parasite some two years previously in India, and had been treated by a pupil of my own with partial success. Portions of the body had come away on five separate occasions. My remedies expelled twenty-five feet of tapeworm, including the finest segments of the neck. In this case the parasite returned at the expiration of seventy days. Curiously enough, this patient was also infested by the larvæ of Anthomyia, of which I obtained four or five specimens. The patient will require further treatment.

CASE C.—P. G. This gentleman is an officer in Her Majesty's Indian army. He consulted me in the early part of February 1875, having been the victim of a tapeworm for a period of about two years. He had been several times treated by an experienced army surgeon, and on one occasion with such a degree of success that no less than fifty-one feet of tapeworm were said to have been dislodged. If that statement be correct there were at least two worms present on the occasion referred to, and one or other of the parasites must have returned or rather grown again. The remedies employed were turpentine, kousso, and male fern. During a very short course of treatment I procured the *head*. Cured.

CONCLUSION.

IF the foregoing 100 cases be analysed and the results compared with those obtained from the 80 cases previously referred to, some slight differences will be noticed. This is owing to the circumstance that in the more extended series I have recorded some 30 new and selected cases, and have added them to 70 of the unselected ones.

Of these 100 cases 95 were genuine, and 5 only were altogether delusive. By genuine cases, I mean that the patients either had a tapeworm when they presented themselves, or they had had one at some time or other previously, from the effects of which they still supposed themselves to be suffering.

Of the 95 genuine cases, 63 victims had contracted the parasite in Great Britain, and 32 abroad. Of these 32 no less than 21 had contracted the worm either in India (including Ceylon) or in other Eastern countries, such as Turkey, Palestine, Persia, and China. The remainder were from the European and American continents.

As regards age, it is instructive to remark that out of the genuine cases 11 victims were under ten years, 22 were under twenty years, and 32 were under thirty years. In other words, more than one-eighth of all the cases were those of young children, one-fourth of the whole series being under twenty years of age, and one-third under thirty years. Of those victims actually harbouring the worm, about two-thirds were males, only one-third being females.

I may observe that of the 95 genuine cases, 16 either sought my opinion only, or obtained merely general and preliminary advice. Altogether there were 22 delusive cases, in 17 of which, however, a worm appears to have previously existed. This leaves only 62 cases actually submitted for treatment. In two of these the results of treatment could not be ascertained.

The 60 cases actually placed under treatment furnish a tolerably fair criterion of the measure of success attending the course of treatment pursued. In two instances only were the results absolutely negative, and in one of these it was owing to the circumstance that the drugs could not be retained by the stomach. In the remaining 58 cases, the entire body or strobile, with more or less of the neck of the worm, was expelled. In no less than 37 instances the *head* of the worm was actually either found by an examination of the evacuation, or was subsequently demonstrated to have been discharged.

In the 21 cases in which the head of the worm was not found, or, as generally happened, was not

so much as even looked for (no opportunities having
been afforded for the necessary explorations), I conjecture that as many as 13 were cures. Consequently
I conclude that, out of the whole 60 cases submitted
to treatment, as many as 50 were ·permanently
cured.   In other words, as I have elsewhere remarked, in all cases where this affection is properly
treated—where the necessary facilities are offered
for exploration, where the physician is thoroughly
well informed as to the natural history and varying
appearances presented by the entozoon in its different stages of growth, where every antecedent
feature in the history of the case is carefully looked
into, or where, in short, there is an exhaustive
knowledge of all the surroundings and possibilities
occurring in any case submitted for diagnosis and
treatment, then, and then only, can that full
measure of success be attainable in the management
of the disorder which in my own experience certainly gives not less than four-fifths of cures in all
cases either actually treated by myself or upon
which I have been called to pronounce and advise
professionally.   As already more than hinted, the
cures· may have amounted to five-sixths of the
whole ; yet, satisfactory as this result undoubtedly
is, I think we have not yet pushed the practical and
scientific management of this disorder to that degree
of perfection which we may yet hope to realise.

Such, at all events, are my facts and deductions.

Whether this estimate of success be above or below the mark, it is given me, thus far, to believe that my efforts to bring about a more rational and complete method of dealing with cases of tapeworm have not been altogether fruitless. One cannot but feel a solid satisfaction in having been the means of curing individual cases of tapeworm which had been quite despaired of, thus relieving many a fellow-creature of the burden of symptoms that were operating to render his or her life well-nigh unendurable.